A fully enclosed appliance supplied to Edinburgh by Dennis in 1933. The pump controls can be seen at the side, and a feature peculiar to this brigade was the large warning light with a green lens mounted above the windscreen.

FIRE ENGINES

Trevor Whitehead

Shire Publications Ltd

CONTENTS

Published in 1997 by Shire Publications Ltd, Cromwell House, Church Street, Princes Risborough, Buckinghamshire HP27 9AA, UK. Copyright © 1981 by Trevor Whitehead. First published 1981; reprinted 1984, 1987, 1991 and 1997. Shire Album 68. ISBN 0 85263 555 9.

Printed in Great Britain by CIT Printing Services, Press Buildings, Merlins Bridge, Haverfordwest, Pembrokeshire SA61 1XF.

ACKNOWLEDGEMENTS

The author acknowledges with thanks the help received from numerous Chief Fire Officers and fire engine manufacturers; curators and staff of museums throughout Britain, and the staff of the British Library in its various locations. Special thanks are due to the following individuals: S. Adamson, M. Burnell, D. Hill, Mr and Mrs D. Riggs, Lieutenant Colonel C. Ross, R. Simpson, R. Vandereyt.

Illustrations are acknowledged as follows: Trevor Whitehead, front cover, pages 3, 7, 9, 10 (bottom), 15, 19 (top), 24 (both), 26 top); I. Scott, pages 1, 18, 25 (bottom), 28 (bottom); R. Vandereyt, pages 2, 19 (bottom), 23; by permission of the Master and Fellows, Magdalene College, Cambridge, page 4; Guildhall Library, City of London, page 6; University of Reading, Institute of Agricultural History and Museum of English Rural Life, page 8 (top); City of Bristol Museum and Art Gallery, page 8 (bottom); Lancashire County Fire Brigade, page 10 (top); Norfolk Museums Service (Bridewell Museum), page 11; Merryweather and Sons Ltd, pages 12, 13 (both), 14 (top), 16 (both); Belfast Transport Museum, page 14 (bottom); Gwynedd Archives Service, page 17; Yorkshire Observer, page 20 (both); D. Hill, pages 21, 30 (both); Carmichaels, page 25 (top); Gwent Fire Brigade, page 26 (bottom); Alex Nicoll and Partners Ltd, page 27; HCB-Angus, pages 28 (top), 29 (top); County of Avon Fire Brigade, page 29 (bottom).

Cover: *Leyland Cub pump purchased by Hinckley Urban District Council about 1940. Now privately owned, it was photographed at a rally at Warwick in 1977.*

Below: *This picture shows the two principal types of ladder used by firemen today; on the right the 45 foot (13.7 m) light alloy one which has largely replaced the wheeled escape, and in the centre a 35 foot (10.7 m) wooden ladder, the type usually carried on every water tender.*

Accurate dating of manual fire engines is often difficult. This one may date from 1680, which, if correct, makes it one of the oldest in existence. It was renovated by junior firemen in 1963.

THE EARLIEST FIRE ENGINES

What is a fire engine? It originated simply as a machine for pumping water, though it has since developed into something considerably more comprehensive. In Britain at the end of the sixteenth century the principal implements for fire fighting consisted of the *siphos*, or squirts, which had been known as far back as the Roman occupation, and leather buckets. The job of using this primitive equipment was left to civic dignitaries and any public-spirited citizens who happened to be available at the time of an outbreak of fire. It was gradually realised that something more effective was required, and in the seventeenth century the first attempts were made to produce a machine capable of quelling the furious element. The main sources of invention and manufacture were located in London. The earliest engines were merely enlarged versions of the old squirts mounted on wheels or small forms of force pump on a sledge or wheels.

The first patent to be issued for a fire-extinguishing machine was granted to Roger Jones in 1625. In this he was acknowledged to have 'perfected a new, profitable and commendable invention, art or skill and way of making and using an engine or instrument artificially wrought with screws and other devices made of copper or brass or other metal for the casting of water, with a spout of copper or brass, into any house, ship or other place taken with fire in such a manner that with the help of ten men to labour at the said engine water may be cast in such abundance and with such violence that it will quench the fire with more ease and speed than five hundred men with the help of buckets and ladders can perform without it.'

In 1634 John Bate described a fire engine which 'will cast the water with a mighty force up to any place you may direct', and in 1652 the city of Exeter bought 'an engine for the quenching of fire' from London. It appears that only two

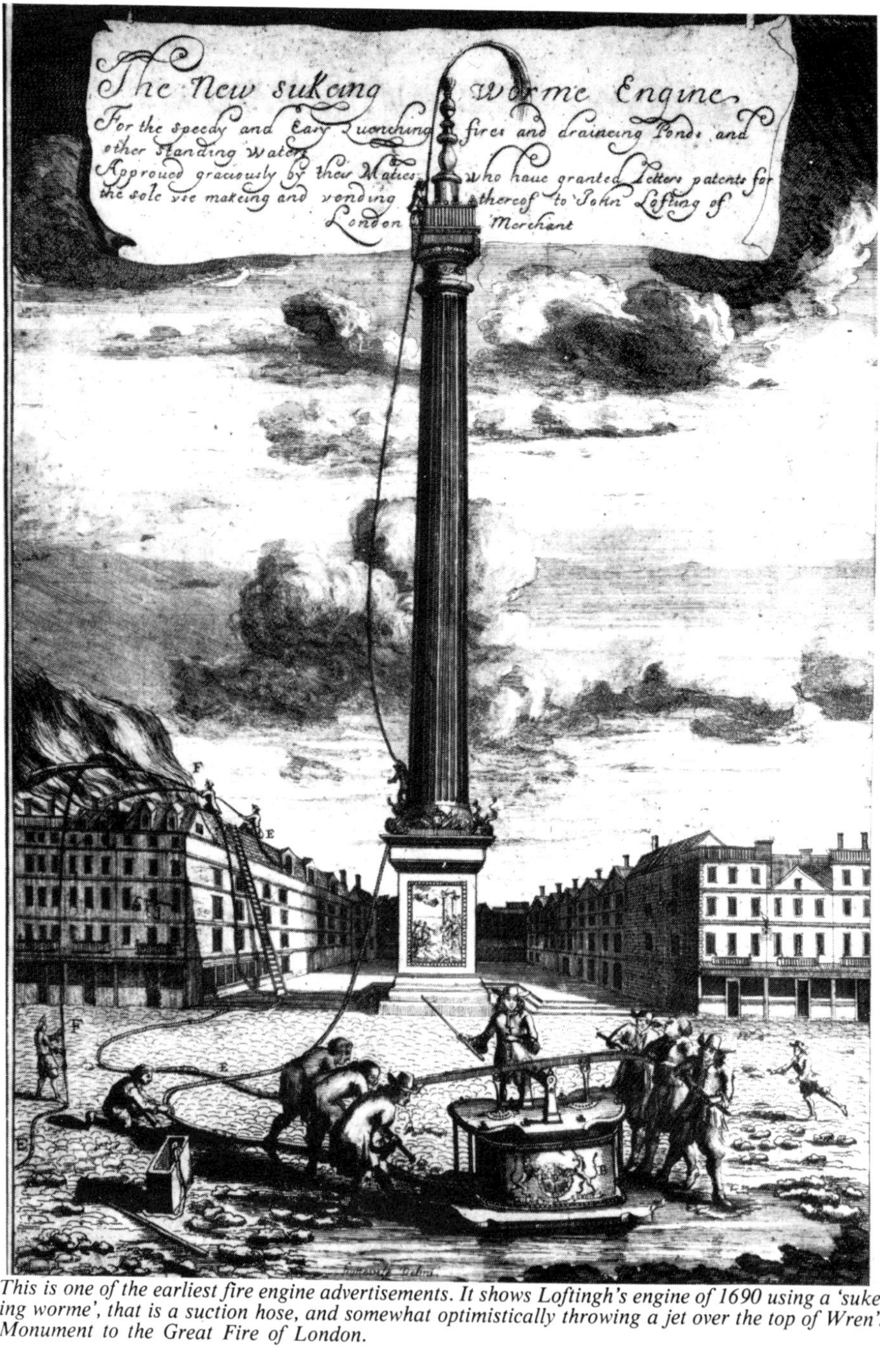

The New sukeing worme Engine
For the speedy and Easy Quenching fires and draineing Ponds and other standing waters
Approued graciously by their Maties: who haue granted Letters patents for the sole vse makeing and vending thereof to John Lofting of London Merchant

This is one of the earliest fire engine advertisements. It shows Loftingh's engine of 1690 using a 'sukeing worme', that is a suction hose, and somewhat optimistically throwing a jet over the top of Wren's Monument to the Great Fire of London.

engines had been built by Roger Jones and the sole rights of his patent lasted only fourteen years. But his basic design was used and improved some years later by William Burroughs, who made 'about three score of these Engines for City and Country'. It is quite clear from municipal records that by 1666 a motley assortment of crude fire engines existed in London and in provincial towns and cities.

At the Great Fire of London it was recorded that 'the engines had no liberty to play for the narrowness of the place and the crowd of people, but some of them were tumbled down in the river, and among the rest that of Clerkenwell esteemed one of the best'. The engine from St James's Church, Clerkenwell, was a Burroughs machine.

The Great Fire of London was such a disaster that men were stirred into action. In 1674 an engine made for Sir Samuel Morland by Isaac Thompson was demonstrated before the king at Whitehall, and in 1676 Strode and Wharton received a patent for their fire engine. Another patent was granted in 1678 to Robert Ledingham. In 1688 several fire engines of a new design were imported from Holland. These were the invention of Jan van der Heijden, superintendent of Amsterdam Fire Brigade, and their main advantage was that they used leather hose instead of the fixed copper branch and nozzle used on the older engines. This enabled the water to be brought to the spot where it was required, perhaps on a rooftop or up stairs inside a building. Previously the jet had to be directed by a man standing on the engine, which therefore had to be placed very close, sometimes too close, to the fire. Quite a number of Dutch engines were imported but their design was soon imitated by English manufacturers. John Loftingh obtained a patent in 1690 for an engine which was very similar to that of van der Heijden; this is not surprising for Loftingh had lived in Amsterdam for about seven years.

Few of the seventeenth-century engines have survived in their original state to be examined by historians. Probably the oldest surviving fire engine to be seen today is in the Museum of London. This one was built for the town of Dunstable in the 1670s by John Keeling, a London fire engine builder, but it is now bereft of its wheels and pumping levers. All that remains is an oval-shaped wooden tub about 4 feet by 3 feet (1200 mm by 900 mm) containing the primitive pump. Another interesting engine of this period, but in a much better state of preservation, was on show at a former carriage museum in Avon. It bears the arms of the Borough of Chipping Sodbury and is stated by Blackstone in his *History of the British Fire Service* to date from 1680. The maker's name is unknown but the construction is more advanced than that of Keeling, and fire engines of this design were being built throughout the following hundred years.

NEWSHAM AND OTHERS

The seventeenth century had been a period of many experiments in building fire-fighting inventions, but it was not until the 1720s that a machine appeared which was to herald the heyday of the manual fire engine. It was built by Richard Newsham, a London pearl button maker, whose premises were situated in the Cloth Fair near Smithfield. In 1721 he was granted a patent for his 'new water engine for the quenching and extinguishing of fires', which was capable of throwing about 110 gallons (500 litres) per minute in a continuous stream.

During the next four years Newsham concentrated on building and improving his fire engines. In 1725 he patented a new design with a long narrow cistern, pumping levers along the sides and treadles for additional pumpers to use their feet on top of the engine. He manufactured six sizes of engine and the largest could deliver about 170 gallons (773 litres) per minute. In the 1725 patent he is no longer described as a pearl button maker, and his own broadsheets and newspaper advertisements use the term 'engineer'. There are still many Newsham engines preserved in museums but the oldest known example is in the village church at Great Wishford in Wiltshire. This was purchased by the churchwardens in 1728 for the sum of £33

Richard Newſham, *of* Cloth-Fair, London, *Engineer*,

MAKES the moſt uſeful and convenient Engines for quenching FIRES, which carry a conſtant Stream with great Force, and yet, at Pleaſure, will water Gardens like ſmall Rain. All impartial Men of Art and Ingenuity will allow this, and the moſt Prejudic'd ceaſe objecting, when they ſee how compleatly the whole Contrivance is adapted to the Uſe intended. He hath play'd theſe Engines before His MAJESTY and the Nobility at St. *James*'s, with ſo general an Approbation, that the largeſt was inſtantly order'd to be left for the Uſe of the Royal Palace aforeſaid: And as a farther Encouragement, (to prevent others from making the like Sort, or any Imitation thereof) His MAJESTY has *ſince* been graciouſly pleas'd to grant him His Second Letters Patent, for the better ſecuring his Property in this, and ſeveral other Inventions for raiſing Water from any Depth, to any Height requir'd. The largeſt Size will go through any Paſſage one Yard wide, in compleat working Order, without taking off, or putting on, *any Thing*; which is not to be parallel'd by *any other Sort* whatſoever: One Man can quickly and eaſily move about the largeſt Size in as little Compaſs of Ground as it takes up to ſtand in, and it is work'd by Hands and Feet, or by Hands only. Thoſe by Suction feed themſelves from a Canal, Pond, or Well, *&c.* or out of their own Ciſterns, as Opportunity offers: They are far leſs liable to Diſorder, much more durable than any extant, and play off large Quantities of Water, at the Diſtances under-mention'd, either from the Engine, or a Leather Pipe, or Pipes, of any Length requir'd; (the Screws all fitting each other) *This* the **cumberſome Squirting-Engines**, which take up four times the Room, cannot perform; nor do they throw one 4th Part of their Water on the Fire, at the like Diſtances, but loſe it by the Way; neither can they uſe a Leather-Pipe with them to much Advantage, whatever Neceſſity may call for. The Four largeſt Sizes go upon Wheels, and the Two others are carried like a Chair. Their Performances are as follow, and their Prizes fix'd very reaſonable, (tho' ſome may think otherwiſe, becauſe his Inventions are ſecur'd to him by Letters Patent) he having a due Regard to the publick Good, as well as his own Profit, both in theſe, and divers other Inventions, for ſeveral Purpoſes, which he has been the Inventor of, either for the Uſefulneſs, or Diverſion of Gentlemen.

Number of Sizes.	What Quantity of Water the Ciſterns hold in Gallons.	Quantity diſcharg'd per Minute in Gallons.	At what Number of Yards Diſtance.	Price without Suction.	Price with Suction.
1ſt.	30	30	26	18 *l.*	20 *l.*
2d.	36	36	28	20	23
3d.	65	65	33	30	35
4th.	90	90	36	35	40
5th.	120	120	40	45	50
6th.	170	170	40	60	70

Machina perfecta eſt, qua non præſtantior ulla
Aſſervare domos, & aquas haurire profundas.

Mutatam cernis naturam: ſurgit in altum
Artibus unda novis; dum flamma coacta recumbit.

3s and it successfully extinguished a fire as late as the mid 1920s.

In the eighteenth century fire engine building became a flourishing business and Newsham had competitors in the trade, but his design was undoubtedly the most popular. There were no public fire brigades as we now know them in existence at that time. Fire fighting, primitive and often disorganised, was undertaken by the parishes and the church was the usual place where the fire engine was kept. The pumping was done by volunteers from the crowd, if any! When insurance companies were formed they organised their own brigades and bought the best available fire engines. At first these insurance fire brigades operated only in the principal cities, but soon they spread to many of the smaller towns.

Newsham continued to sell his engines in Britain and overseas until his death in 1743. The business then passed to his son Lawrence and after his death to his widow and his cousin George Ragg, the firm then becoming Newsham and Ragg. This carried on until 1765 and may then have been taken over by John Bristow, though this is not certain. Bristow's engines were almost identical to those made by Newsham and there are a number extant. Another manufacturer well established by the middle of the century was Hadley of Long Acre, predecessor of the famous Merryweather firm.

The typical eighteenth-century fire engine comprised a cistern of oak, mahogany or other hardwood mounted on four wheels. Two single-acting pump barrels were placed in the cistern and an air vessel was connected to the delivery outlet to ensure a continuous discharge. On some models the water had to be poured into the cistern using leather buckets, but most engines had a suction inlet for hose so that the cistern could be filled either by suction or by buckets. Although Newsham's second design introduced pumping levers placed laterally other makers retained the former type of transverse levers. There were also variations in the position of the outlet, the long branch on top of the engine remaining in favour for many years after leather delivery hose had been invented. These engines were provided with a

LEFT: *After Newsham had patented his second design he advertised his engines in glowing terms. This broadsheet was issued in 1728 and shows the capabilities claimed for the different sizes, and their prices.*

RIGHT: *The oldest known manual engine by Richard Newsham. This historic fire engine, complete with leather hose and copper fittings, is preserved in the church of St Giles in Great Wishford. It was bought by the churchwardens in 1728.*

ABOVE: *The firm of Tilley flourished in London from 1820 to 1851. This example of their large horse-drawn manual engine was built in 1823 for Folkingham in Lincolnshire.*
BELOW: *A typical manual engine of the latter half of the nineteenth century. The pumping levers are in the folded position for travelling. This was built by Shand Mason in 1867 for the Imperial Fire Insurance Company's brigade in Bristol and is now in the city's Industrial Museum.*

draghandle but sometimes, particularly in rural areas where long distances had to be travelled, a special horse-drawn wagon was built to transport the fire engine.

In the latter half of the eighteenth century modifications and improvements were made. These included larger spoke wheels instead of solid wood, the attachment of the front wheels to a forecarriage making the engine more manoeuvrable, delivery outlets at the front or sides of the cistern, and the provision of shafts so that the engine could be pulled by a horse. An important alteration was the placing of the valves apart from the pump barrels, thus making them easily accessible. This improvement was patented in 1792 by Charles Simpkin, who later joined the Hadley firm. The foot treadles invented by Newsham were copied by a few other makers, notably by Bristow, and they continued to be made until the early years of the nineteenth century.

A quite different type of pump was invented in 1785 by Joseph Bramah. The mechanism in engines by Newsham and others consisted of iron segments and chains attached to piston rods, but Bramah used a new kind of mechanism, in which, in his own words, 'the advantages principally arise from the piston's having its motion round a centre, in a rotary direction, instead of reciprocating in a straight cylinder'. In 1793 he patented another design, but despite being extremely well made these Bramah engines do not appear to have been commercially successful.

At the start of the nineteenth century many towns possessed fire engines but few people were trained to operate them. There is some doubt as to which town was the first to organise its own paid body of men for fire fighting, but it was probably Beverley in Humberside in 1726. The first city to have a municipal fire brigade was Edinburgh in 1824, but many towns continued to rely on insurance fire brigades for another fifty or sixty years.

One of the few extant engines built to Joseph Bramah's first design of a semi-rotary pump. It was made in 1791 for the Earl of Strafford and is displayed in the Fire Service College.

ABOVE: *A rare example of a manual with a semi-rotary pump, built by Rowntree in 1799 for the Earl of Derby's estate. Note the similarity to the earlier pump made by Bramah.*

BELOW: *This Merryweather steamer was purchased in 1885 by a private landowner in Berkshire. It was delivered four days after the order was placed — a contrast to modern delivery times!*

A fine specimen of a large Shand Mason steamer, preserved in Norwich Bridewell Museum. It was bought in 1881 for £700 by Colman, the mustard manufacturers, for their works fire brigade. The quick steam-raising apparatus can be seen attached to the funnel.

THE ERA OF STEAM

The first fire engine to have the pump worked by steam was invented in 1829 by Braithwaite and Ericsson of London. The vertical boiler at the rear supplied steam to two horizontal cylinders which enabled the engine to throw 150 gallons (682 litres) per minute to a height of 90 feet (27 m). Although this machine was apparently well designed, weighing 2¼ tons (2,286 kg) and equipped for horse draught, it met with determined opposition. Some people were afraid that they would lose the beer money paid to those who worked the manual pumps. Others contended that the water mains could not adequately supply such an engine, that it was too heavy, too expensive and would cause excessive water damage. During the next four years only four further engines were built, the idea was dropped, and the populace continued to be protected from fire by the old manual engines.

Before the early nineteenth century the emphasis in fire fighting was on saving

In the 1880s Merryweather introduced their 'Metropolitan' single cylinder steamer, designed originally for the London brigade. The pumping machinery was fitted horizontally at the side of the boiler and a large flywheel ensured even motion.

property, with little serious thought being given to the saving of human life. In 1836 the Royal Society for the Protection of Life from Fire — an organisation supported by voluntary contributions — was formed with the intention of providing fire escapes. These consisted of a main ladder about 35 feet (10.7 m) long mounted on a pair of large wheels. A fly ladder was attached to the top of the main ladder and this could be swung up by means of ropes to give an additional 10 feet (3.05 m). Fire escapes of this type were used for many years until replaced by an improved pattern using telescopic ladders. This item of equipment was carried on fire engines up to the 1970s, since when it has largely been replaced by a 45 foot (13.7 m) light alloy ladder.

It was halfway through the nineteenth century before the steam-driven pump reappeared. The first successful steamer was built in 1858 by Shand and Mason, a London firm which had been in the fire engine business since 1774. The older firm of Merryweather produced their first steamer in 1861 and this too was a success. At the International Exhibition held in Hyde Park in 1862 the latest mechanical inventions were shown to the public and great interest was aroused by three steam fire engines, two by Shand and Mason and one by Merryweather. The following year a three-day trial of fire engines was held at the Crystal Palace in south London and this event attracted no less than ten steamers, seven of British manufacture and three from America. The first prize for large engines went to Merryweather and happily this engine, though not with its original boiler, is preserved in the Science Museum in London. In the small engine class Shand and Mason were awarded first prize and they rapidly built up an export trade as well as supplying engines to many British brigades. Although a number of other engineering companies built one or two, the market for steam fire engines from the 1860s to the First World War was almost entirely supplied by Merryweather and by Shand Mason. By 1878 Merryweather had built five hundred. The only other company successfully to market a steamer was Wil-

ABOVE: *The 'County Council' pattern was made in four sizes, the largest of which could pump 300 gallons (1,363 litres) per minute. The engine was placed vertically in front of the boiler and the two fuel boxes at the rear were a prominent feature.*
BELOW: *The new 'Greenwich' manual was patented by Merryweather in 1890. Outwardly similar to their 1851 design, it incorporated mechanical improvements claimed to make it more powerful without increase of weight.*

ABOVE: *Chemical engines became popular in the late 1890s. On this horse-drawn Merryweather machine two 20 gallon (90 litre) cylinders were connected to 180 feet (55 m) of hose, and a 35 foot (10.5 m) ladder was carried on top.*

BELOW: *A 'Metropolitan' double vertical engine. This rare example of a steamer built by Rose and Company of Manchester was used by Belfast Fire Brigade from 1892 to 1911.*

A Shand Mason steamer built in 1896 for the Metropolitan Fire Brigade (London). It was sold to Sturminster Newton, Dorset, in 1912 and attended its last fire in 1933.

liam Rose of Manchester.

The steamer soon became the front-line fire engine of all the newly established municipal fire brigades but the manual engine was far from being superseded. During the nineteenth century the manual was gradually transformed from a small manhandled pumping machine into a large fire engine capable of transporting the crew and all their equipment and drawn by one or two horses. It was still dependent on muscle power to operate the pumping levers, and the number of men required to work one of the larger engines varied between twenty-two and forty-six. A new design invented by Moses Merryweather in 1851 and known as the 'London Brigade' manual was widely used. Other firms such as Shand Mason, Tylor, Rose and Warner sold almost identical machines in large numbers.

The Victorian era was a time of rapid growth in the fire engine business. Many steamers and manuals were supplied to royalty, the nobility and gentry to protect their huge houses and vast estates. When a town took delivery of a new steamer it was an occasion for much festivity, and the new appliance was often named *Victoria* or *Albert* or perhaps given the name of the mayor or the chairman of the fire brigade committee. There was immense local pride in the fire brigade and the firemen were regarded as heroes. A steamer at full gallop with smoke pouring from the funnel and the crew's brass helmets gleaming must have been a thrilling sight and sound.

ABOVE: *Merryweather 'Fire King' supplied to Brighton Fire Brigade in 1903. The steam engine provided power to drive the rear wheels as well as to work the pump, which had an output of 400 gallons (1,818 litres) per minute.*
BELOW: *The first petrol motor appliance carrying a chemical engine, hose reel and wheeled escape. Built by Merryweather in 1903 for Tottenham, this machine joined a motor 'Fire King' in the first urban fire station to be designed without any provision for horses.*

Leyland fire engines were always popular, and this machine with Rees-Roturbo pump at the rear and solid rubber tyres was the typical design of the early 1920s.

THE MOTOR TAKES OVER

At the beginning of the twentieth century the internal combustion engine, already successful in cars, was being used increasingly for commercial vehicles. Its main advantage over other forms of propulsion was the speed with which full power was obtainable. But fire engine manufacturers still put their faith in steam and many refinements had been incorporated in the latest engines, including quick steam-raising apparatus and variable expansion gear to make more economical use of the steam power. Some steamers were converted to oil fuelling as an alternative to coal, coke or wood. In 1899 Merryweather designed a self-propelled steamer and by 1906 they had supplied eighteen to the home market and twelve for export.

Special appliances to supplement the basic pumps were being introduced in many fire brigades: horse-drawn escape tenders to transport the escape ladders which had formerly been pushed along the streets by hand; self-supporting mechanical turntable ladders able to reach greater heights than the escapes; chemical engines for fighting small outbreaks, consisting of a 35 or 60 gallon (159 or 273 litre) tank actuated on the same principle as a soda-acid extinguisher.

The first self-propelled petrol motor fire engine to be used by a public fire brigade was built by Merryweather for Tottenham in 1903. This had a 60 gallon chemical engine with 180 feet (54.9 m) of small-diameter hose and it carried an escape. The following year Finchley fire brigade took

In 1928 Leeds purchased this Merryweather turntable ladder, which had a wooden 85 foot (25.9 m) ladder and Hatfield pump. In addition to the monitor attached to the top of the ladder, an uncommon feature of this appliance was the larger one mounted on the pump.

delivery of the first petrol motor fire engine to have its own 250 gallon (1,136.5 litre) per minute pump operated by the road engine, in addition to a chemical engine and escape. This too was made by Merryweather, and it may now be seen in the Science Museum in London.

In the early years of the twentieth century there were experiments with battery-electric and petrol-electric fire engines but they never became widely used. Many oddities were produced by attempts to convert steamers to motor traction but the petrol motor soon proved its superiority although the early motors were far from trouble-free.

In 1908 and 1909 two firms which were to become world-renowned for their appliances entered the fire engine business. Dennis Brothers of Guildford built a motor pump for Bradford fire brigade in 1908 and Leyland of Lancashire built one for Dublin fire brigade in 1909. Both were excellent products and orders immediately began to flow in to both firms. 1909 was also the year in which Bristol purchased a Shand Mason self-propelled steamer, but probably the first city fire brigade to be entirely equipped with motors was Belfast in 1914.

But throughout Britain the phasing out of horses was spread over a period of more than twenty years. Fire horses were much loved animals and their passing was regretted by the general public as well as by many firemen.

In 1922 Shand Mason was absorbed by its long established rival Merryweather, and three firms dominated the market — Dennis, Leyland and Merryweather. Among other manufacturers of fine fire engines may be mentioned John Morris of Manchester, Halley of Glasgow and Tilling-Stevens of Maidstone.

When motor appliances first appeared the bodywork remained remarkably like that of the earlier manual engines, with the crew seated on either side of what was virtually a box for containing the hose and small items of equipment. This gave no protection from the weather and was dangerous, as men were thrown off the machine when cornering at speed. This Braidwood type body (named after the famous Superintendent of the London Fire Engine Establishment) continued in production until after the Second World War, but the first fully enclosed limousine pumps were commissioned as early as 1931, a Merryweather going to Edinburgh and a Dennis to Darlington.

By 1930 solid tyres were being replaced by pneumatic ones, giving a safer and more

ABOVE: *This Leyland/Metz 100 foot (30 m) turntable ladder, built in 1935, originally belonged to Oldham Fire Brigade. Many of these German ladders were imported before the war.*
BELOW: *In the 1930s many fire engines were built with limousine bodywork. This emergency tender with a Dennis Ace chassis was purchased by Sheffield in 1934.*

ABOVE: *A Home Office escape-carrying unit of the early 1940s on a Ford chassis. It had a 130 gallon (590 litre) water tank and a special extensible tow bar was fitted to prevent the trailer pump from fouling the wheels of the escape.*
BELOW: *A mobile dam unit of the National Fire Service. The sides have been let down to show the canvas dam and suction hose connected to the light pump at the rear.*

During the war years Merryweather produced a 60 foot (18 m) hand-operated turntable ladder using the Austin K4 chassis. Having served with the NFS since 1943, this was stationed after the war at Bridgwater in Somerset.

comfortable ride. The turntable ladder became very popular, first with wooden ladders of 75 feet (22.8 m) or 85 feet (25.9 m) and in the 1930s with steel ones of 100 feet (30 m). There was only one long-established British manufacturer of these large appliances — Merryweather, but many brigades favoured the German ladders built by Magirus or Metz and mounted on Dennis or Leyland chassis. The sale of trailer pumps to some of the smaller brigades began in the 1920s, and production was greatly increased in 1938 when war seemed imminent and the Auxiliary Fire Service was formed to augment the regular fire brigades.

THE WAR YEARS

At the outbreak of war in September 1939 there were about 1,400 local authority fire brigades in England and Wales and about 185 in Scotland. Except in the few cases where neighbouring authorities had made an agreement for mutual assistance, each brigade operated strictly within the limits of its own local authority boundary. There was an incredible variation in the equipment, ranging from the most up-to-date motor vehicles to, in some towns and villages, steamers and manual engines! When the real horror of the blitz began in the autumn of 1940 the local authority fire brigades and the Auxiliary Fire Service were called upon to deal with fires of a size and complexity hitherto unknown. Bitter experience clearly demonstrated that the peace-time organisation and methods of the Fire Service were totally unsuitable for war conditions, and in August 1941 the National Fire Service was created. This absorbed the AFS and all the brigades formerly administered by local authorities.

One item of fire-fighting equipment which proved most valuable during the war years was the trailer pump. This could be manoeuvred over piles of rubble and past bomb craters where a full-size fire engine could not go. Trailer pumps were manufactured by about fourteen firms including Dennis, Coventry Climax, Beresford and Sigmund. The smallest was known as the wheelbarrow and had a single-cylinder engine of 300 cc, which delivered about 50 gallons (227 litres) per minute. The more powerful ones were classified as light, medium or large, the last having a four-cylinder engine of from 15 to 30 horsepower capable of delivering 500 gallons (2,273 litres) per minute.

The government had ordered the production of vast numbers of trailer pumps for the Auxiliary Fire Service and it was assumed that they would be towed by private cars, taxis and any other vehicles available. This did not prove entirely successful, and later a Home Office designed towing vehicle was produced on the Austin K2 chassis with a 24 horsepower six-cylinder engine. It was fitted with a steel body with strengthened roof for protection against shell splinters, ample locker space and a waterproof apron to close the rear of the machine. The auxiliary towing vehicle, or ATV, with a large trailer pump became the most widely used appliance of the war years, proving versatile and reliable.

Fire engines were so urgently required in such large numbers that standard government specifications were issued and machines built on different chassis by different coachbuilders. Two heavy pumping appliances were made and these consisted of a pump capable of delivering 700 gallons (3,182 litres) or 1,100 gallons (5,000 litres) per minute mounted on a Bedford, Ford or Austin chassis, with lockers for hose and small gear and gallows for carrying ladders. These were used to fight the enormous blitz fires and to relay water over long distances to other pumps. Escape-carrying units with a pump mounted in front of the radiator were produced, and many different types of fireboat were built to protect the docks and riverside risks in cities and estuarial towns. All National Fire Service appliances were painted grey, in contrast to the bright red of the pre-war fire engines.

In addition to the front-line pumping appliances many other special kinds were built to deal with the conditions of war. One of the greatest problems posed by air raids was water shortage. At the very time when there were simultaneous large fires the water mains were destroyed by high explosive bombs, and so an emergency hose-laying lorry was built with a capacity of 6,000 feet (1,830 m) of 3½ inch (89 mm) rubber-lined hose. Mobile dam units, consisting of a flat-bed lorry carrying a collapsible steel-frame dam holding 500 or 1,000 gallons (2,273 or 4,546 litres), also helped to alleviate the water shortage. Other emergency appliances included control units and field telephone vans for ensuring adequate communications; petrol tankers, mobile repair vans and breakdown lorries to keep the fire-fighting appliances in action; canteen vans and mobile kitchens to cater for the many people at work for long periods.

Turntable ladders could no longer be imported from Germany, so the whole burden of production fell on Merryweather. Additional premises were taken over by this firm in order to cope with the output of 100 foot (30 m) and 60 foot (18 m) turntable ladders. The former were mounted on Dennis or Leyland chassis, the ladders being identical to those manufactured before the war, but the latter were an innovation. The 60 foot ladders were hand-operated and mounted on an Austin chassis, being designed to be used by any fire crew as distinct from the specially qualified operators of the 100 foot mechanical ladders.

Although the NFS was never tested to its full potential, having been formed after the air raids had become less severe, the emergency appliances produced from 1941 onwards served their purpose well.

In the 1950s Miles of Cheltenham were building water tenders with all metal bodies. This one was on a Dennis F8 chassis for Shropshire Fire Brigade.

THE SECOND HALF OF THE TWENTIETH CENTURY

After the war had ended the Fire Service had to make do with its emergency-type appliances, and many were adapted to a peacetime role and remained in service for some years. The two largest pre-war manufacturers, Merryweather and Dennis, gradually resumed production and introduced new designs, but Leyland decided not to continue in the fire engine business. In 1948 the National Fire Service was disbanded and the responsibility for providing a fire service was restored to local authorities, but with two important differences from before the war — a smaller number of authorities and an element of central control retained by the Home Office. Fire engines built since then conform to certain minimum standards relating to such matters as weight distribution, engine performance, road performance and amount of water carried. This ensures a high all-round standard but leaves manufacturers and individual brigades free to design and develop appliances to suit particular needs.

One fire engine which developed from wartime experience was the water tender. Its forerunner was the mobile dam unit carrying enough water for initial fire fighting and towing a trailer pump. The post-war machine was known as a Type A water tender but this was soon replaced by the Type B, which has its own built-in pump, thus avoiding the disadvantages of towing. In 1947 Dennis produced the first in a new range of appliances called the F series, but these were open-bodied and similar to pre-war machines. In 1949 they

Dennis F8, known as the 'Ulster'. This appliance, only 6 feet 6 inches (1981 mm) wide, was designed for work in confined areas or rural districts. Northern Ireland Fire Authority commissioned thirty of these between 1952 and 1954.

introduced a completely new limousine design, the F7, which had a 1,000 gallon (4,546 litre) per minute pump and a Rolls-Royce petrol engine. This had a 162 inch (4,115 mm) wheelbase but this was found to be too long, and in 1950 the F12 appeared, mechanically an identical machine but with a 150 inch (3,810 mm) wheelbase.

Merryweather were selling a similar appliance but using an AEC diesel engine, and large numbers of this and the Dennis F12 were supplied to replace wartime and pre-war machines. As the demand for fire engines increased other companies entered the market, notably HCB of Southampton and Carmichael of Worcester, as well as a number of smaller firms. Suitably modified commercial chassis such as Bedford, Ford, Albion and Commer were used, with Bedford being the most popular. Between 1958 and 1963 Leyland briefly re-entered the fire engine market with their Firemaster chassis but only ten of these were built.

Two appliances used by Middlesex Fire Brigade: on the left a Dennis F12 pump escape of 1953, and on the right a Dennis F7 pump of 1950.

ABOVE: *The Commer was a popular chassis in the 1950s and many water tenders were built to this design by Carmichael. The visual warning at that time was two orange flashers above the windscreen.*
BELOW: *Leyland 'Firemaster' pump escape commissioned in 1960, and now preserved in Glasgow Transport Museum. The doors enclosing the front-mounted pump can be seen below the windscreen.*

ABOVE: *Another Leyland 'Firemaster'. This was built in 1962 and combined the functions of pumping appliance, emergency tender and salvage tender, carrying a wide range of specialised equipment.*

BELOW: *This was the first hydraulic platform to be bought by a British fire brigade. It could reach a height of 65 feet (19.8 m) and had a 100 gallon (455 litre) water tank, hose reel and rear-mounted pump. Built on a Commer chassis, it was in service from 1963 to 1979.*

The Land Rover has been the basis of many small fire appliances. This one was built in 1969 by HCB-Angus. Roof gallows carry ladder and suction hose and the pump has an output of 400 gallons (1,816 litres) per minute.

There was a period of controversy concerning the best colour for a fire engine. White and yellow appliances were in vogue for some years but red is once again the standard colour, though lockers are often left unpainted embossed aluminium.

This book has been confined to outlining the evolution of the basic fire engine — a machine for pumping water — but as the complexity of fire fighting and rescue operations has increased so ancillary appliances have been devised. In 1963 Monmouthshire Fire Brigade ordered the first 'Snorkel', an hydraulic platform which could be used for rescue or for directing a large jet of water from a height of 65 feet (19.8 m). As the articulated booms could be directed over and down behind buildings this type of appliance soon became widely used, and units are available reaching heights of 77 feet (23.5 m), 91 feet (27.8 m) or 103 feet (31.5 m). For some years it was thought that these platforms would supersede turntable ladders but it is now recognised that each appliance has its own role in fire fighting and rescue operations.

Two other post-war developments have been the gradual replacement of wooden wheeled escapes by ladders made of light aluminium alloy and the introduction of high expansion foam equipment for combating fires in ships' holds and other confined spaces. Other appliances used by fire brigades today include rescue tenders and decontamination units for dealing with road accidents and with chemicals or radio-active materials that have caught fire or been spilled, foam tenders for use at petrol or oil fires and control units which serve as mobile headquarters at large incidents. In the late 1970s the need for better protection of the firemen in case of road accidents resulted in the Crew Safety Vehicle designed by HCB-Angus. The cab is capable of withstanding the impact of a very severe crash. The modern fire engine, radio-equipped and manned by a highly trained crew, is an efficient, sophisticated and costly machine, far removed from the 'commendable invention' of 1625. But whereas the motor fire engine has been with us only since Edwardian times the manual fire engine in its various forms was in use in Britain for almost three hundred years.

ABOVE: *Water tender built by HCB-Angus on an AEC Mercury chassis in 1970. This one has a combined siren and flasher in addition to the traditional bell.*
BELOW: *The Scoosher was designed in 1969 by Glasgow Fire Service. It is a hydraulic arm which extends to 45 feet (13.7 m) and has at its tip a heat detector, a piercing implement and a nozzle for either a jet or a spray. This Dodge/Carmichael entered service in 1973.*

ABOVE: *A light fire appliance built by HCB in 1967 and based on a Ford Transit van. At the rear is a Coventry Climax pump, and a reel of first aid hose is mounted above a 60 gallon (273 litre) water tank.*
BELOW: *Rescue tenders are a vital part of a modern fire brigade's fleet. The wide range of equipment includes winches, jacks, saws, bolt cutters, axes and foam branches. Halogen floodlights on a hydraulic mast and portable lights are powered by the vehicle's own generator.*

ABOVE: *Firemen using the cage of a Simon hydraulic platform as a vantage point to fight a fire in stacks of waste paper. Two large jacks on each side give this heavy appliance a firm base when the booms are elevated.*

BELOW: *This Dodge chassis was widely used in the 1970s. The wheels visible on the roof are used in conjunction with a portable pump, which is carried in one of the lockers.*

FURTHER READING

Blackstone, G.V. *A History of the British Fire Service*. Routledge & Kegan Paul, 1957.
Bonner, R.F. *Manchester Fire Brigade*. Archive Publications, 1988.
Burgess-Wise, David. *Fire Engines and Fire-Fighting*. Octopus Books, 1977.
Creighton, John. *Fire Engines of Yesterday*. Ian Henry Publications, 1984.
Dudley, Ernest. *Chance and the Fire Horses*. Harvill Press, 1972.
Firebrace, Sir Aylmer. *Fire Service Memories*. Andrew Melrose, 1948.
Goodenough, Simon. *Fire: The Story of the Fire Engine*. Orbis Publishing, 1985.
Green-Hughes, Evan. *A History of Firefighting*. Moorland Publishing, 1979.
HMSO. *Fire Fighting Appliances*. Catalogue of the collection in the Science Museum, 1969.
HMSO. *Manual of Firemanship. Part 5: Ladders and Appliances*. Revised 1993.
Holliss, B., and Thompson, J. *The Green Machine*. Athena Books, 1995.
Holloway, Sally. *London's Noble Fire Brigades 1833-1904*. Cassell & Co, 1971.
Holloway, Sally. *Courage High!* HMSO, 1992.
Ingram, Arthur. *Fire Engines in Colour*. Blandford Press, 1973.
Ingram, Arthur. *A History of Fire-Fighting and Equipment*. New English Library, 1978.
Jackson, W. Eric. *London's Fire Brigades*. Longmans, Green & Co, 1966.
Mallett, Janette. *Fire Engines of the World*. Osprey Publishing, 1982.
Miller, Denis. *A Source Book of Fire Engines*. Ward Lock, 1983.
Nicholls, Arthur. *Going to Blazes*. Robert Hale, 1978.
Pennington, Roger. *British Fire Engine Heritage*. Osprey Automotive, 1994.
Perry, G.A. *Fire and the Fire Service*. Blandford Press, 1972.
Rickards, Maurice. *The World Fights Fire*. Longman, 1971.
Smith, Peter. *Go to Blazes*. Milestone Publications, 1986.
Thomas, Bernard. *Fire-Fighting in Maidstone*. Phillimore, 1976.
Thorpe, Peter. *Moonraker Firemen*. Wiltshire Library and Museum Service, 1979.
Vanderveen, Bart H. *Fire-Fighting Vehicles 1840-1950*. Frederick Warne, 1972.
Vince, John. *Fire-marks*. Shire, 1973 (reprint 1989).
Wallington, Neil. *Images of Fire*. David & Charles, 1989.
Wallington, N., and Holloway, S. *Fire and Rescue*. Patrick Stephens, 1994.
Wright, Brian. *Firefighting Equipment*. Shire, 1989.
Wright, Brian. *Firemen's Uniform*. Shire, 1991.

THE FIRE BRIGADE SOCIETY

This society was formed in 1963 to promote and foster interest in fire brigades and their appliances. Members are enabled to enlarge and develop their knowledge of all aspects of the fire service, past and present. Membership is open to anyone interested in this subject, and full details may be obtained from: M. T. Williams, 115 Barnwood Avenue, Gloucester GL4 7AG.

PLACES TO VISIT

The following list is a brief selection from the many museums which include one or more fire engines in their collections. Those marked * are devoted exclusively to fire-fighting appliances.

Birmingham Museum of Science and Industry, Newhall Street, Birmingham B3 1RZ. Telephone: 0121-235 1661.

Bodmin Museum, Mount Folly, Bodmin, Cornwall PL31 2HQ. Telephone: 01208 74159.

Bridewell Museum, Bridewell Alley, St Andrew's Street, Norwich, Norfolk NR2 1AQ. Telephone: 01603 667228.

Bristol Industrial Museum, Prince's Wharf, Prince Street, Bristol BS1 4RN. Telephone: 0117-925 1470.

British Commercial Vehicle Museum, King Street, Leyland, Lancashire PR5 1LE. Telephone: 01772 451011.

The British Engineerium, Nevill Road, Hove, East Sussex BN3 7QA. Telephone: 01273 559583 or 554070.

Central Museum and Art Gallery, Guildhall Road, Northampton NN1 1DP. Telephone: 01604 233500.

Dorset County Museum, High West Street, Dorchester, Dorset DT1 1XA. Telephone: 01305 262735.

**Fire Museum*, 101-109 West Bar, Sheffield, South Yorkshire. Telephone: 0114-249 1999.

Guildford Museum, Castle Arch, Quarry Street, Guildford, Surrey GU1 3SX. Telephone: 01483 444750.

Guildhall Museum, High Street, Rochester, Kent ME1 1QU. Telephone: 01634 848717.

Herefordshire Waterworks Museum, Broomy Hill, Hereford. Telephone: 01432 761733.

Horsham Museum, 9 The Causeway, Horsham, West Sussex RH12 1HE. Telephone: 01403 254959.

Liverpool Museum, William Brown Street, Liverpool L3 8EN. Telephone: 0151-207 0001.

Long Shop Museum, Main Street, Leiston, Suffolk IP16 4ES. Telephone: 01728 832189.

Lostwithiel Museum, Fore Street, Lostwithiel, Cornwall PL22 0AS. Telephone: 01208 872513.

**Museum of Fire*, Lothian and Borders Fire Brigade, Lauriston Place, Edinburgh EH3 9DE. Telephone: 0131-228 2401.

Museum of London, London Wall, London EC2Y 5HN. Telephone: 0171-600 3699.

Museum of Transport, Kelvin Hall, Bunhouse Road, Glasgow G3 8PZ. Telephone: 0141-287 2720.

Royal Albert Memorial Museum, Queen Street, Exeter, Devon EX4 3RX. Telephone: 01392 265858.

Salisbury and South Wiltshire Museum, The King's House, 65 The Close, Salisbury, Wiltshire SP1 2EN. Telephone: 01722 332151.

Science Museum, Exhibition Road, South Kensington, London SW7 2DD. Telephone: 0171-938 8000.

Shaftesbury Local History Museum, Gold Hill, Shaftesbury, Dorset. Telephone: 01747 852157.

South Molton Museum, Guildhall, South Molton, Devon EX36 3AB. Telephone: 01769 572951.

Staffordshire County Museum, Shugborough, Stafford ST17 0XB. Telephone: 01889 881388.

Swansea Maritime and Industrial Museum, Museum Square, Maritime Quarter, Swansea, West Glamorgan SA1 1SN. Telephone: 01792 650351 or 470371.

Welsh Industrial and Maritime Museum, Bute Street, Cardiff CF1 6AN. Telephone: 01222 481919.

Whitstable Museum, Oxford Street, Whitstable, Kent CT5 1DB. Telephone: 01227 276988.